The Zero-Point Paradox

Villacres

ISBN: 978-1-7374905-1-7

DEDICATION

This book is dedicated to those that look at the skies.

CONTENTS

INTRODUCTION

To begin, let's look at B.K. Ridley's words from his book *Time, Space, and Things*. There, he points to the ideal piece of matter in its ideal shape and form. He creates this fictional object called the *utopium* ball. This is a perfectly spherical object with uniform density. "Physics is about the simple things of the universe." He says, a rock is too complicated in shape, the rough surface has equally complicated aerodynamics, too noisy to unravel our laws of kinematics. The five platonic solids are not uniform enough, either. So in general, we use the utopium ball: an imaginary object that helps us understand how ballistics works. Perfectly round, perfectly smooth, and as large as we want it—be it a planet, a star, an asteroid, or an apple. (Not black holes, however. Only objects in Newtonian physics.) In Hawking's book, *A Brief History of Time*, he remarks on how solids behave like solids, liquids like liquids, and gases like gases, independent of the material the matter is made of. In other words, Ridley's utopium ball works perfectly well for our models in physics.

THE ZERO-POINT PARADOX

A brief look at Newton's model of mass in 2*D* space:

|Figure 1|

If we model this physical situation, we get:

| Figure 2 |

Here, we have masses M_1 and M_2 with a distance r between the $0D$ points. They are the reference points to the center of masses of each object.

Let's illustrate the gravitational field of M_1 in a fixed position and M_2 as our test point. This acceleration field felt by M_2 around M_1 is given by:

$$\mathbf{a} = G\frac{M_1\mathbf{x}}{|\mathbf{x}|^3}.$$

The acceleration vector **a** depends on the position vector **x** of the object with mass M_2. G is the gravitational constant and M_1 the mass of the fixed object. With an arbitrary position vector, we see the vector field around M_1:

|Figure 3|

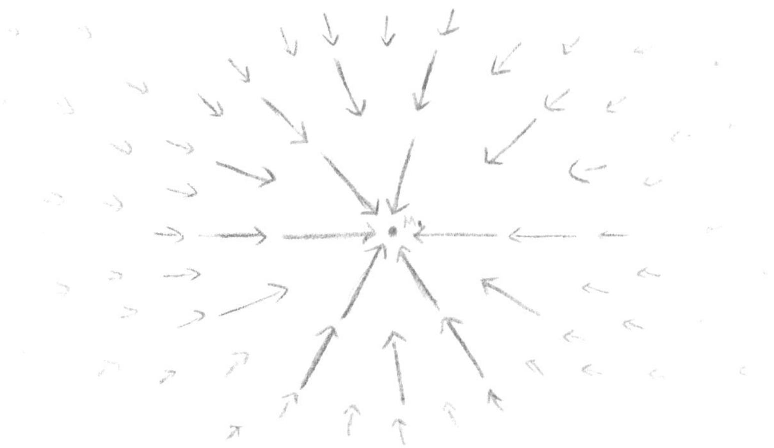

This is all known and well understood. And here is where I introduce what I call the "**zero-point paradox**:"

As $|\mathbf{x}| \to 0$, $|\mathbf{a}| \to \infty$. However, at the very center where $|\mathbf{x}| = 0$, $|\mathbf{a}| = 0$.

To make the paradox a bit clearer: you weigh more the closer you are to the center of mass (COM) due to the increase in acceleration(gravity). If you're twice as close to the COM, you're four times your current weight due to the squared inverse relationship. But then, at the exact position of the COM, you are weightless.

Let's see what Newton's Shell Method says about this. According to the Shell Method, when in a position inside the mass, the "shells" of mass above do not affect the acceleration of the object. In other words, the acceleration field would not go to infinity because as we enter the object itself, the value of the mass decreases in magnitude. The field looks more like this:

|Figure 4|

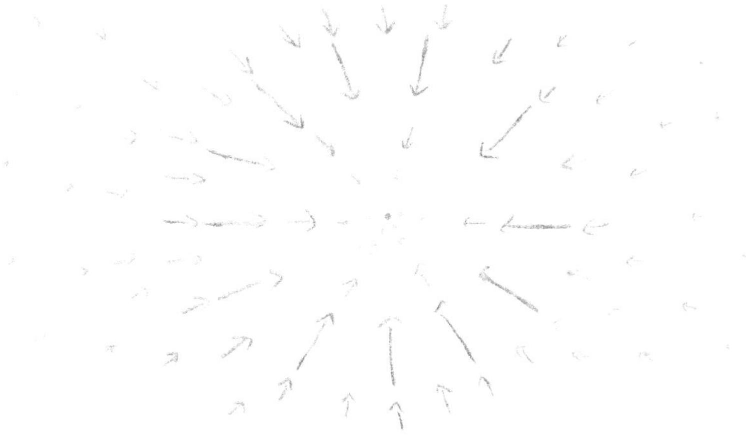

In essence, we get

$$\mathbf{a}(m_1) = G\frac{m_1\mathbf{x}}{|\mathbf{x}|^3}.$$

Here, $m_1 < M_1$ and the mass m_1 decreases monotonically to 0 as $|\mathbf{x}| \to 0$. A question of density comes up from this perspective. It is apparent that density increases, but the possibility that density could remain constant or decrease after a certain distance remains unclear.

Also, with the help of the Shell Method, the Center Of Mass is shown to be a reference point to the mass of the object. This is just to highlight the difference between Center Of Mass (COM) and the mass at the center. It is known that a COM can lie outside its body, so there's no surprise there.

The COM is a reference point that changes in value depending on the "shell" and mass it encompasses. The mass the at center is the physical piece of mass that lies at the center.

It seems Newton's Laws may be incomplete when it comes to making a model of a "heavenly body" spinning with a stable rotation. I make this assertion because of the Zero-Point Paradox. Having large and increasing magnitudes of gravity next to a point of no gravity is peculiar. And as we approach the COM of the object, density clearly increases. But there's a question of whether density could potentially decrease. This possibility, that after a certain distance the gravitational field diminishes, will be investigated.

I will explain my model next.

2-DIMENSIONS

Let's look at a *2D* model in *2D* space.

|Figure 5|

This disk has mass M_1 (I'll be reusing some names for masses throughout) and is perfectly circular so that the COM is at the center. Let M_1 be a playdough-like mass, so it is not entirely solid and not entirely liquid (i.e. non-rigid). This will be our utopium ball.

The center point-mass, *m* (the **physical piece** of mass), that rests at the COM (a **reference point in 0D** that varies in its value depending on the mass in a shell) has a gravitational acceleration of 0.

Since $|\mathbf{a}| = 0$ there, then $|\mathbf{F}| = m|\mathbf{a}| = 0$.

Claim: By Newton's 3rd Law, if the mass m at the center of M_1 has a force $|\mathbf{F}| = 0$, then m is a separate object that does not affect M_1.

Just to reiterate, point-mass m could have a large mass value (like a singularity) or have no mass at all, M_1 is unaffected by it. Realistically it's unlikely there's a singularity at the COM, but I'm only highlighting a peculiarity in Newton's Equations.

|Figure 6|

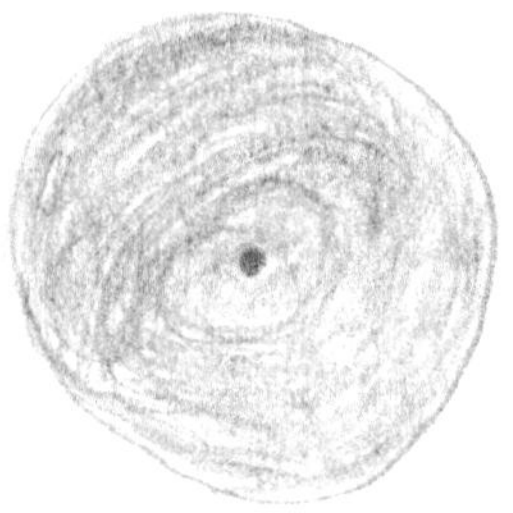

In this figure, the point-mass m is a $0D$ point. Removing this mass would not affect the mass of M_1 because m has no action (gravitational action) on M_1. So let's suppose we have removed that mass at the **0D point**, we can think of this "hole" as a **new COM: a 1D circle with radius of 0**. A $1D$ circle with radius 0 is essentially equivalent to that $0D$ point we removed.

However, the reason I want to refer to the point as a circle is because of the implication that the circle can be expanded, perhaps if energy were directed at it. Then, we would have a **region** (not just a point) of space where the gravitational potential is 0 within. Before we look into how that energy could be applied, let's first see what it would be like to have a **region** inside a 1*D* circle COM with a **non-zero** radius.

Let's call this circle the *COM*(1*D*) for "center of mass in 1-Dimension." Also, we will refer to the 0*D* point as *COM*(0*D*), to keep things clear.

2-DIMENSIONS-EXAMPLE

Let's begin with this example. We will look at a new similar piece of mass, M_1, in the shape of a thin CD. Time is considered frozen and we start the clock after analyzing both the outside and inside regions. Here we will loosely call the inner circle the $COM(1D)$.

So this is the $2D$ physical object M_1 in $2D$ space:

| Figure 7 |

First, we draw the $COM(0D)$ of M_1:

|Figure 8|

Then we add on top the $COM(1D)$ of M_1 with its radius, $r > 0$. For now, we will use the inner wall of M_1 as our $COM(1D)$.

|Figure 9|

And to repeat, we are using this new object $COM(1D)$ loosely until it's established more concretely.

We suppose this CD is not rotating. From outside the $COM(1D)$, M_1 is tending towards the $COM(0D)$.

Let's suppose we have a mass m_β with some velocity, $\mathbf{v}$, inside the region $[COM(0D), COM(1D)]$. According to Newton's Shell Method, m_β does not feel the mass M_1 around it because m_β is inside this region(shell). m_β is not tending towards the $COM(0D)$. It is weightless relative to M_1, experiencing no gravitational potential of M_1 .

m_β will eventually hit the inner wall of M_1 with a kinetic energy of $\frac{1}{2}m_\beta|\mathbf{v}|^2$. Let's suppose that m_β bounces off in an inelastic collision so that kinetic energy is conserved (and $\mathbf{v}$ is conserved.) This means we (could) have m_β bouncing off the walls inside of M_1 like so:

|Figure 10|

m_β is tending towards the $COM(1D)$, but is only doing so at discrete points along this circle.

Therefore, this model is unstable. The radius of $COM(1D)$ will go to 0. m_β is not providing enough kinetic energy evenly throughout the $COM(1D)$ to hold this "CD" open. The CD will shrink to a disk with no hole (except at the center point-mass) after time runs forward. For this model to be stable, we would want M_1 to be tending towards the $COM(0D)$ and m_β tending towards the $COM(1D)$ evenly at all points, and with some extra conditions we will get to.

* * *

In the second example, we will start the same, but this time mass M_1 will be rotating at a **constant** velocity, $\mathbf{v_0}$. We forgo the physical mass and go straight to the COM model, with the addition of the velocity:

|Figure 11|

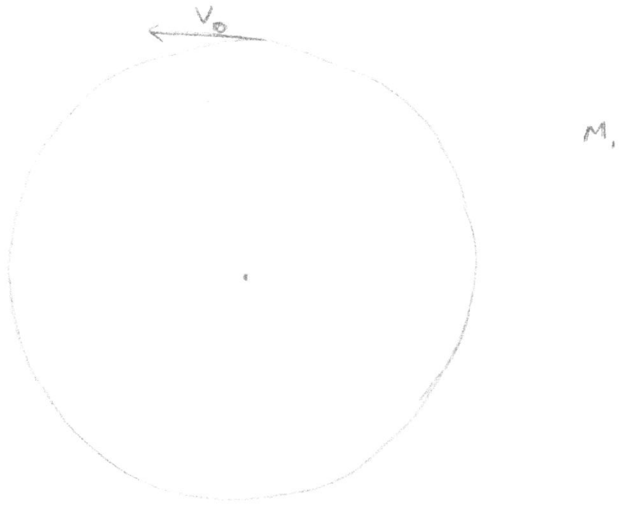

Again, we have m_β with velocity, $\mathbf{v}$, and it will eventually hit the inner wall of M_1 like before. This time, since the wall is rotating with tangent velocity $\mathbf{v_0}$, m_β will either speed up (if it's slower than $\mathbf{v_0}$) or it will slow down (if it's faster than $\mathbf{v_0}$) on impact because the velocity vector of m_β reflects relative to the tangent velocity of the rotating $COM(1D)$: (If m_β was going in a counter direction, we see the direction change until they are both in the same orientation.)

|Figure 12|

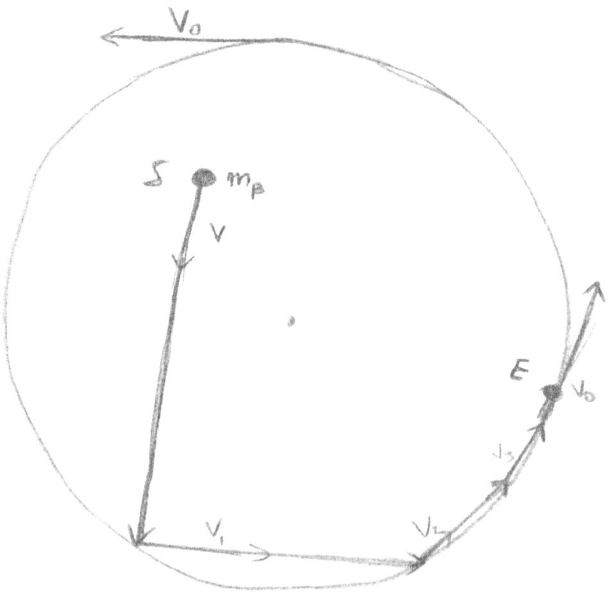

Let's mark the start of m_β with **S**, and the end of m_β with **E**. m_β is tending towards the $COM(1D)$ and is being spread throughout the $COM(1D)$ from within. m_β is being pushed by the inner wall of M_1. Let's picture m_β as a collection of many marbles, bouncing away inside with their own velocities until they hit the inner walls of this CD. They hold the circle open with the mass at a final velocity $\mathbf{v_0}$. What I will depict is the trajectories of a few marbles. But it's apparent that with enough small masses, they would spread throughout the whole inside of the walls.

|Figure 13|

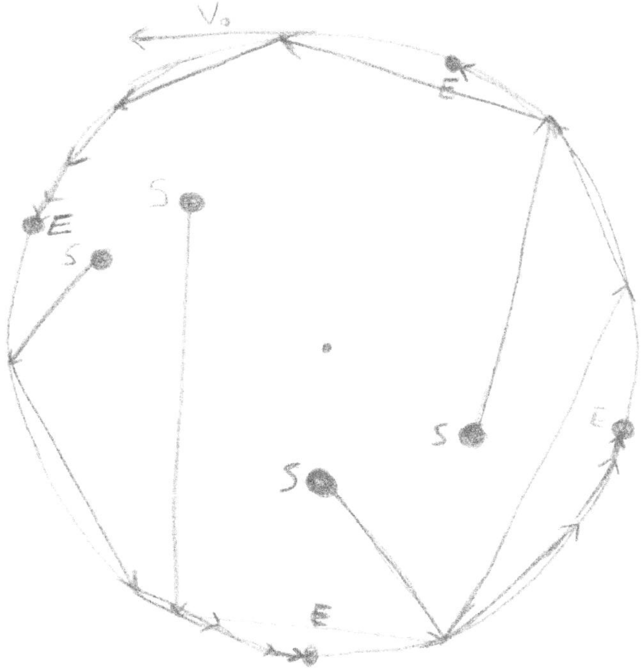

This is my stable model. It is the rotation of M_1 that allows for a stable $COM(1D)$. From outside the $COM(1D)$, M_1 is tending towards its $COM(0D)$ as well as towards the $COM(1D)$, which preserves the general idea of COM. And from within (in the region where the gravitational potential of M_1 is 0 due to Shell's Method) we have m_β being pushed against the inner wall of the $COM(1D)$ with a final kinetic energy $\frac{1}{2}m_\beta|\mathbf{v_0}|^2$.

Later, we will look more at the conditions for m_β.

3-DIMENSIONS

To make a stable model in $3D$, let's use our stable $2D$ model as our basis. We will use the COM model in a stable rotation with some rotating mass in 2D (with the idea that some inside mass "m_β" is being pushed within):

|Figure 14|

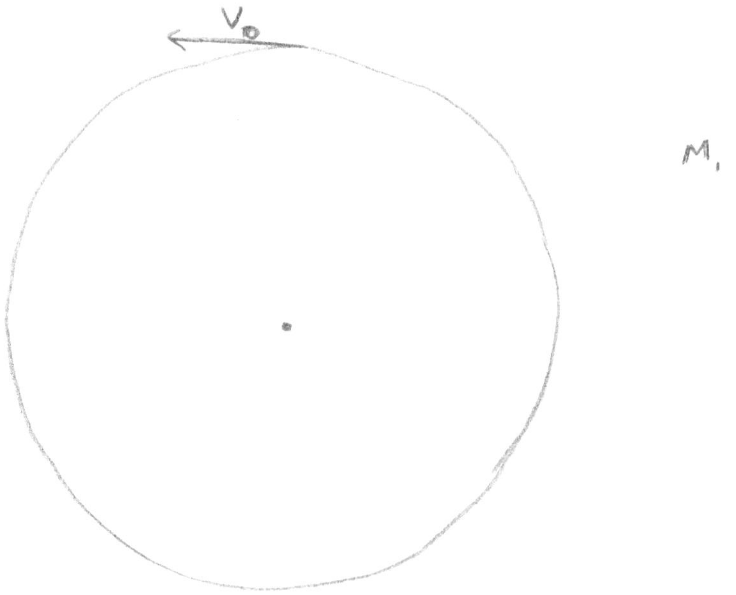

Let's stack these $COM(1D)$ "sheets" into the $3D$ space as follows:

|Figure 15|

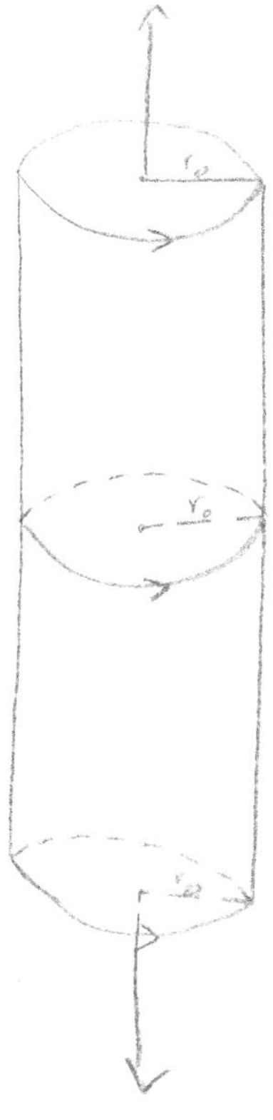

Stacking them like this seems to never end. It would require incredibly large values for the mass M_1 outside and, twice that mass, m_β inside to hold this system together. The height for such a structure would be cosmically large, so we could consider this quasi-stable. We can think about such a structure later since they are not modeling the rotating systems in question. Stars and planets are not long cylinders, but spherical.

Using some intuition, however, we can stack them such that. as we move above (and below) the "original circle" $COM(1D)$, we preserve the distance between the circumference of each new $COM(1D)$ and the center of the first circle to that of the radius of the first circle r_0. In other words, $r_i(h)^2 = r_0^2 - |h|^2$ as $|h| \rightarrow r_0$ where $r_i(h)$ is the radius of the next "sheet" at the distance h above and below the original. r_0 is the radius of the original $COM(1D)$ disk. This gives us the following:

|Figure 16|

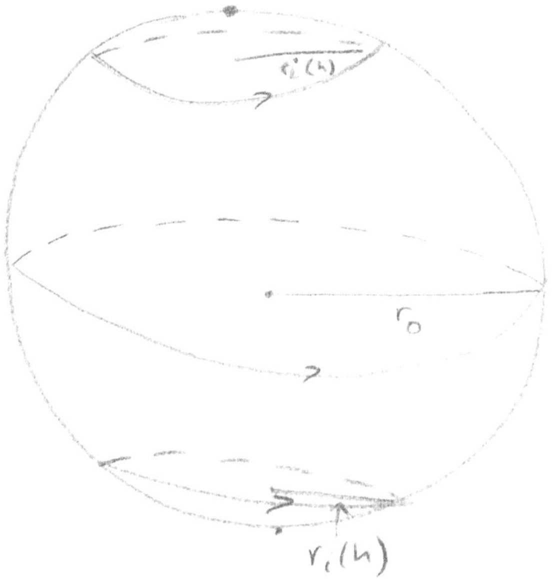

This model has a limit to its height and is built of stable $2D$ sheets stacked on top of each other. Also, we now see the three COM objects: a point or the $COM(0D)$, a circle or the $COM(1D)$, and a cylindrical sheet or the $COM(2D)$. The $COM(2D)$ is much like the cylinder in Figure 15, except with the cylinder in the shape of a sphere. We built this model such that we had M_1 be all layers of mass outside the circles and m_β be all layers of mass inside the circles.

ENERGY AND STABILITY

So far, I have presented a model for COM's in 3 dimensions, but I have vaguely established if Center Of Mass truly is a proper name for these objects in the 3 dimensions.

Let's look at the Potential Energy of a mass M_1, like in Figure 7, rotating with velocity $\mathbf{v}$ in $2D$ space(since $3D$ is a natural extension of $2D$ objects as we saw in the previous chapter.) First, we calculate the acceleration of gravity at radius r, the radius of the circle $COM(1D)$ from the point $COM(0D)$. m_2 is some test point for our acceleration field equation of M_1 at r.

$$m_2 g = G\frac{M_1 m_2}{r^2} \text{ so that } g = G\frac{M_1}{r^2}$$

This gives us a potential of

$$PE_{out} = M_1 g d = M_1 G\frac{M_1}{r^2} r = G\frac{M_1^2}{r}.$$

This is the PE of M_1 being a distance $d = r$ from $COM(0D)$. One reason we get M_1^2 in this equation is because we extend the properties of

Center Of Mass in $0D$ to the new Center of Mass objects that I have introduced. In other words, the $COM(0D)$, $COM(1D)$, $COM(2D)$ all act as if the mass M_1 lie completely on these objects.

m_β has kinetic energy from being pushed within M_1. We get that

$$KE_{in} = \frac{1}{2} m_\beta |\mathbf{v}|^2$$

For this to be a stable $COM(1D)$, we would want

$$PE_{out} = KE_{in}$$

So we get

$$G\frac{M_1^2}{r} = \frac{1}{2} m_\beta |\mathbf{v}|^2$$

Let's divide by r,

$$G\frac{M_1^2}{r^2} = \frac{1}{2} m_\beta \frac{|\mathbf{v}|^2}{r}$$

So we get that

$$M_1 = \frac{1}{2} m_\beta, \text{ or } 2M_1 = m_\beta.$$

This means that for a stable system (and since mass can't tell the difference between gravitational acceleration or centripetal acceleration, i.e. $g = \frac{|\mathbf{v}|^2}{r}$ is a valid statement in physics), we would need that m_β be twice the mass of M_1.

Equal and opposite reactions implies the following: The action of gravity's pull on mass has an equal and opposite reaction of a push on mass.

ROTATING EXPERIMENTS AND OBSERVATIONS

Rotation of mass is really interesting in the perspective of dimensions. For example, within a ***1D*-space** **(i.e a line),** rotation of mass is impossible. A mass could not rotate since rotation involves at least two dimensions. We are supposing mass exists in these spaces.

* * *

Rotation in ***2D*-space** **(i.e a plane)**, also places limits on a rotating mass. As we saw, a stable rotation in $2D$ involved the $COM(1D)$, a circle. But this circle separates the $2D$ space it inhabits: an inside and outside of the $COM(1D)$, like in Figure 9. This limits the system by separating the space. Perhaps we can imagine this by looking at rigid objects spinning on an axis. Because of the rigidity, the inside and outside are separate.

The Cone and Tennis Ball

During my time at San Francisco's Exploratorium, they had a device made with a cone blowing air to a ball like so:

|Figure 17|

You could angle the cone as well, but the ball would stay fixed at that "zero-point" like so:

|Figure 18|

This experiment highlights how our intuition can betray us about objects falling down (or being pushed away) and shows the forces at play: centripetal acceleration and the force of gravity.

The Apple and Air Compressor

This is another experiment that can help us understand rotation in 2*D* (or just in general): the spinning apple using an air compressor.

1) Take an air compressor and turn it on, angling it up to about 60 degrees from the ground

|Figure 19|

2) Aim an apple to fall at the "zero-point" of the blast of air so that the apple gains rotation in place like so:

| Figure 20 |

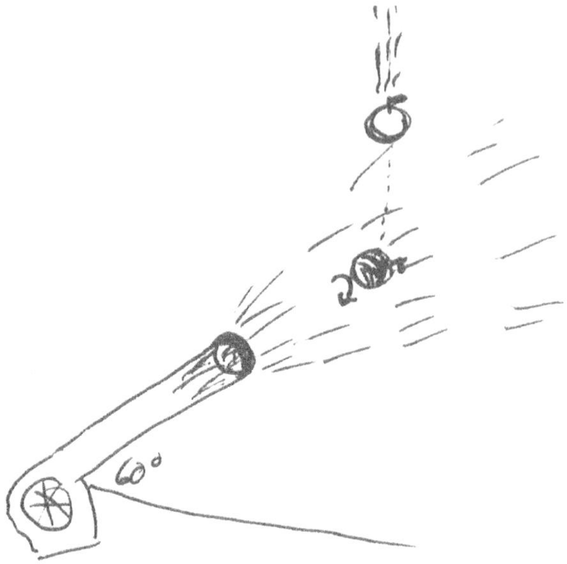

As the apple begins to spin and settle on an axis, you will notice the speed of rotation increase steadily. A planar slice of this rigid apple, perpendicular to its axis of rotation, is very much like a mass spinning in 2*D* space. Think of this one slice as we see the experiment play out. This model says of this slice: the $COM(1D)$ is expanding because of the increasing kinetic energy and some of the mass that was originally weighing the object down is now being pushed (the energy provided by the air compressor). Then, when the kinetic energy from within is too great (because the $COM(1D)$ is getting too large for the body it's in) to hold the apple together, it suddenly explodes outwardly, perpendicular from the axis of rotation.

| Figure 21 |

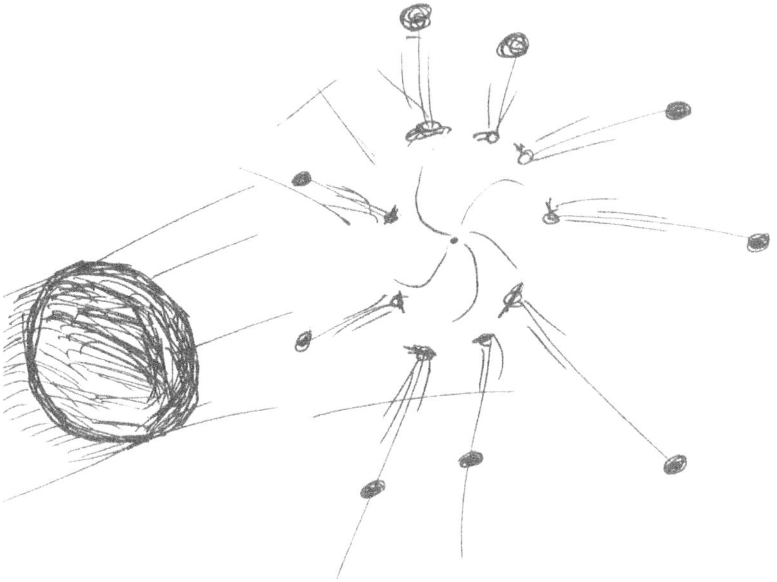

Truly, this experiment is better seen in the many videos available of it.

The Wheel

Adding a spin to an object creates a stability in the axis of the spin. This is seen in footballs when players add spin for more accuracy. Gyroscopes are used to stabilize rocket paths. When the rocket turns from its path, the gyroscopes have a restoring force from the angular momentum that pushes the rocket in the opposite direction of its deviation. The previous examples show how angular momentum predictably stabilizes the COM's. In The Wheel experiment, the idea can also be seen in a not-so-predictable context.

Here, we have a bicycle wheel with spokes like so:

|Figure 20|

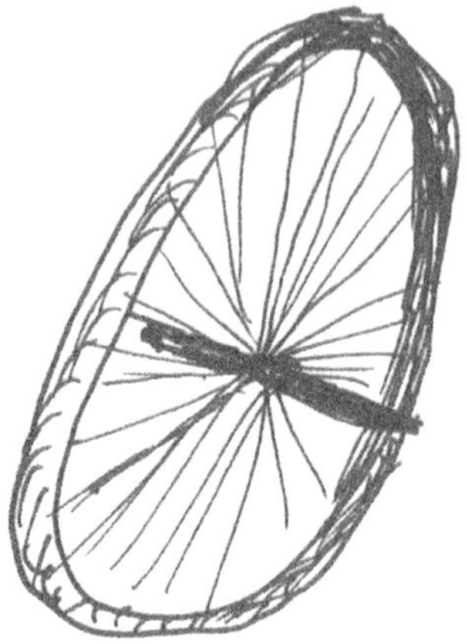

We will hang the wheel by fixing one of the spokes to a chain held by the ceiling, and see what happens.:

|Figure 21|

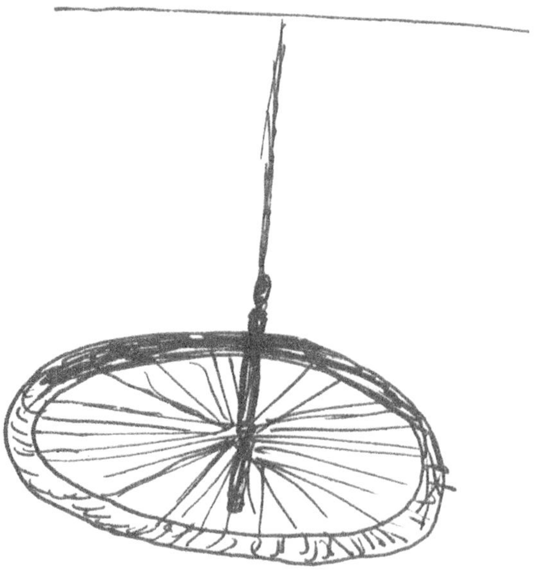

Without rotation, this wheel predictably falls.

Now let's add spin to the wheel and hang it the same way:

|Figure 22|

The wheel will hold in this position for longer than intuitively expected. The spin (i.e. the angular momentum) stabilizes the object. The COMs are also stable for a while. As soon as we place the wheel on the chain, the free spoke experiences a pull from Earth's gravity and this adds torque to the system. The result is that the wheel rotates around the chain's axis as it continues to spin. Because of friction, the wheel loses spin and eventually falls back down. Still, the next part of this demonstration is more intriguing

—adding weight to that free spoke after setting the wheel on the chain:

| Figure 23 |

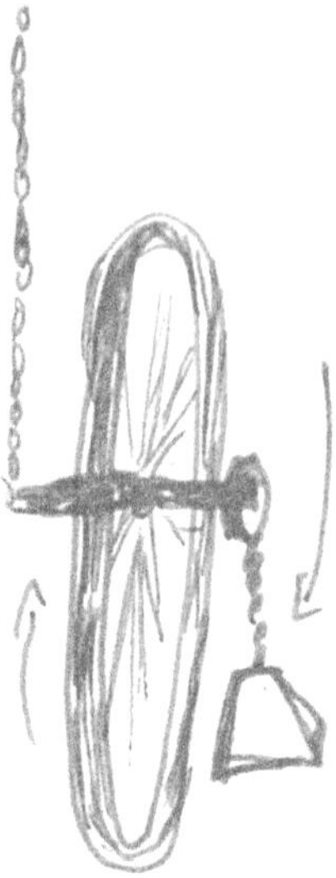

The torque increases because the pull downward increases, and the wheel rotates faster around the chain's axis until the wheel looses its spin. This experiment is also great to see in video or in person. It's a demonstration that shows how a stable(ish) rotating system can hold mass up.

* * *

Observations of rotation in **_3D_-space:** By the process of stacking "circles" on top of each other, the top and bottom circles of the $COM(2D)$ had a radius of 0 by construction. This allows us to assume that an open "cylinder" is the correct way to view our $2D$ COM sheet instead of what looks very much like a closed "sphere". This is important because, unlike in our $2D$-space, our $COM(2D)$ does not completely separate the inside and outside $3D$-space. A crucial feature for this model.

Suppose for a moment that these COM objects are valid extensions of Newton's logic for center of mass and gravity. What are the implications? Planets, stars, black holes: these objects' structures (among other things) are in question with this model. It would be interesting to apply this model with general relativity and see the behavior of such objects as black holes, but our masses and our rotational velocities here are in Newtonian context.

Water Droplet in Microgravity

There's a video of a rotating water droplet in a microgravity environment. In this experiment, the rotating water ball is held together by the weak hydrogen bonds. They spin the water droplet and perform various experiments: they drop different objects into the water such as salt, rocks, and tea leaves. Due to their varying densities, each material behaved differently in the water. But the structure was clear. Air bubbles accumulated at the center axis of rotation of the water. And the tea leaves were driven to the surface of the "ball." This is very much like the structures I have described. I recommend you view the video of this experiment as well. (*Rotating Sphere of Water in Microgravity, 2008.,* https://www.youtube.com/watch?v=BxyfiBGCwhQ)

A Planet-Satellite Origin

This next example will be partly a thought experiment. What does this model with the COMs say within our planetary system? Can we find any evidence?

Let's look at our Moon, a potential piece of evidence. The contending theory is that the Moon is a result of a crash from long ago between Earth and another planet, Theia. The theory argues: after the crash, dust and space material formed a disk around Earth, and over millions of years, the gravity of Earth caused the Moon to accrete from the materials of the crash. This **Accretion Theory** would imply a composition from the Moon of the objects that crashed. However, in the Apollo missions, we discovered that the Moon's composition was identical to Earth's inner mantle. An impact that large would definitely leave evidence of its impact with another planet, but we did not find it.

Furthermore, the accretion theory doesn't explain the high angular momentum system of the Earth and Moon. We do not have simulations or explanations to show the Moon gathering such high angular momentum after accretion. Though this accretion theory is at the forefront, it is also highly unlikely. (*Impact Origin of the Moon*, Erik Ausphaug, 2014: https://

www.researchgate.net/publication/

280321193_Impact_Origin_of_the_Moon)

Another theory of the origin of the Moon is that from fission. The idea is that a large piece of mass gained kinetic energy from our early Earth, with its molten body and faster speed of rotation, and was launched into orbit from Earth. This is the mostly likely idea due to the composition of the Moon, but the theory lacked the mechanism to show how.

This fission idea requires an energy far too great to burst through the pressures of gravity. So this idea was not taken seriously, despite the evidence of the Moon being part of this fission process.

Using my theory, however, we can intuit the following sequence:

|Figure 24|

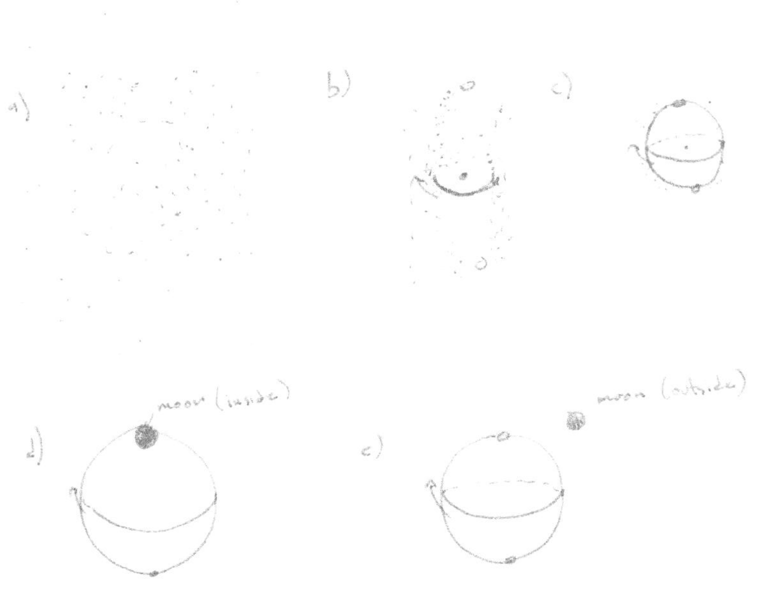

The hot early Earth is volatile, excess energy inside the system could send a part of m_β out from inside as shown. It could explain the high angular momentum, the fact the Earth and Moon are practically the same age, and how the Moon has a composition similar to the inside mantle of the Earth. Without having to push through the crust of Earth because the North and South poles make the least resistance in the hot early Earth. The hot early Moon is launched, already spinning, from the centrifugal force.

This sequence of events is similar to another theory made from graduate students Simon Lock and Sarah Stewart. They called it Synestia. The difference between Synestia and my theory is where the kinetic energy comes from to send the Moon from the Earth's inside. Synestia, unlike my theory, still required another planet Theia to crash into Earth for the energy. The simulation provided by the graduate students, however, show the same shapes and actions this theory intuits, as shown in Figure 24.

(*Where did the Moon come from? A new theory*, Sarah Stewart, 13 March 2019: https://www.ted.com/talks/sarah_t_stewart_where_did_the_moon_come_from_a_new_theory/)

Seismology and the The Shadow Zone,

The core of the Earth is a geological theory in direct contradiction to this model. Geologists will point to seismology for further evidence against such a structure in planets, let alone stars. I must concede academically to seismologists on the topic, this perspective is less from studying seismology and is more theoretical in approach. However, I must make a serious attempt to see if perhaps there is evidence and an explanation for the claims presented here.

From my understanding, the model built from seismology makes the following assumptions: pressure and density increase all the way to the center. If we look at Figure 3, then the magnitude of the gravitational field correlates with this assumption. But as was shown in the "Zero-Point Paradox" this is not the whole story. Newton's Shell Method, Figure 4, allows us to draw a more accurate picture of gravitational pull that no longer has that same correlation. So the assumption of density and pressure should not be taken for granted, if true.

In this book we assume it's not true and have come up with the model as we have. In this model with a rotating mass M_1 (and m_β) in $3D$ space, we assume that M_1 does feels the increasing pressure and density until some distance where we have m_β that does not feel anything but the push from M_1 inside the $COM(2D)$ region.

If the m_β-region is much less dense relative to the M_1-region, we can have this seismic model where the dot represents the earthquake's origin:

| Figure 25 |

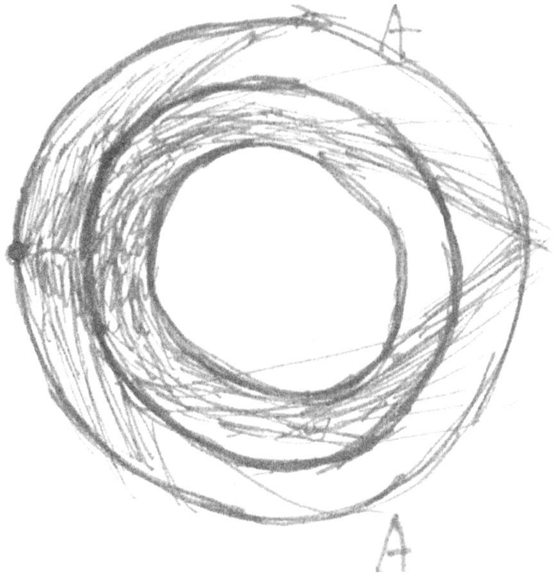

Because the m_β-region is much less dense, earthquake waves would bend much more dramatically as they travel through this region, allowing the wave to pass to the other side.

The Shadow Zone is an area on the surface of the Earth where waves do not hit. In the current theory, are blocked or refracted by the dense core as they travel inside the Earth. I have labelled this region as *A*. Still, inside this shadow zone, some waves do end up in there regardless. How? This fact led geologists to conclude that the core must have an inner and outer core. The interaction of those layers with the waves explained the sparse earthquakes that made it through. But the shadow zone could also be explained without a core at all. With the two regions as shown, we see the shadow zone, as well as the sparse waves that make it through.

In addition, the longer routes the waves take would explain why seismic waves **appear to slow down** as they travel "through the core" to opposite side of the wave's origin. This is significant because, theoretically, the waves should speed up in dense regions and arrive sooner than when measured.

The New Core Paradox

To further highlight a discrepancy between the Earth-Core model and real observations, we must understand the New-Core Paradox.

Simply put, geological studies of Earth established a core to be highly unlikely, or at the very least, admits there is little understanding in **how** cores are created naturally for moons and planets. Specifically, it shows that the magnetic field of Earth is older by a billion years than the core itself. This shows that there may be a different mechanism for how our magnetic field is generated. At the very least, it shows that we know very little how our core formed, if at all.

The concept of the core is such a contested idea even today. The heat and pressure make it difficult for any magnet to retain a field. The inner and outer core convection theory has flaws. Many of these flaws are pointed out by nuclear chemist Dr. J Marvin Hendon in his paper, *Reasons Why Geomagnetic Field Generation is Physically Impossible in Earth's Fluid Core.* However, his proposition to justify the inner and outer core (i.e. the magnetic field phenomenon) is a georeactor sub-shell surrounding a georeactor sub-core. That's at least 4 layers of "cores" that exist to justify the cores function: creating a magnetic field and diverting seismic waves. Clearly, the Earth-Core concept is not as solid as is taught.

(*Reasons Why Geomagnetic Field Generation is Physically Impossible in Earth's Fluid Core,* Advances in Social Sciences Research Journal – Vol. 8, No. 5, Dr. Herndon, 25 May 2021: http://www.nuclearplanet.com/geomag4.pdf)

Dark Matter

The Core-Cusp Problem is a discrepancy, in part, between the proposed density at the center of galaxies and the measured density. Models predict an increasingly dense center for galaxies. This is due to the power-law assumption of gravity, based on the assumption of ever-increasing densities on a smaller regions. But, observations show that galaxies have a flat central dark matter density profile. This is in agreement with my model: my model implies that in a rotating system, there is a limit where density does not continue to increase, but it stays constant or decreases. This is a possible explanation to the observations of galaxies in the Core-Cusp Problem.

This model also fits with the observations of Dark Matter in general. Dark Matter is described as matter that does not show up with our measurements of gravity and mass, but is still in much more abundance than regular matter, and practically invisible. Here, Dark Matter takes the form of m_β in all the rotating systems of the cosmos. Whether this takes into account all of dark matter or a large portion of it, is another question.

The Sun

If this model is true for a planet, it must also be true for our sun. Keep in mind that the sun holds 98% of all the mass in the solar system. By the logic of my model, a cloud of hydrogen gas would be split into two parts. 1/3 of the mass would be outside some radius with gravity pulling it in. 2/3 of the mass would be inside that radius being pushed outwardly. In other words, we would expect that a cloud of hydrogen gas "loses" mass from before and after the birth of a star. Is this what we see?

Through NASA's Hubble and Spitzer telescopes, and many years of observing the birth of many stars, they found this is exactly the case. The mass of the cloud of hydrogen and the mass of the stars created from the cloud have been measured to lose 70% of their matter. So, from the state of being a cloud of gas, to the state of being a star, the amount of hydrogen mass lost is 70%. My model predicts these observations, especially if we consider that $PE_{out} \leq KE_{in}$ instead of $PE_{out} = KE_{in}$.

(*Hubble Shows Torrential Outflows from Infant Stars May Not Stop Them from Growing*, March 18, 2021: https://www.nasa.gov/feature/goddard/2021/hubble-shows-torrential-outflows-from-infant-stars-may-not-stop-them-from-growing)

Density Reasonability

Suppose we have two spheres of radius 1 with the **same mass** and each has **uniform density,** except one is hollowed out. The hollowness radius is given by distance from the center, r. At what value r will the density of one ball be double the other ball? That happens when the volume gets cut in half. So,

$$V_1 = \frac{4}{3}\pi(1)^3$$

$$V_2 = \frac{4}{3}\pi(1)^3 - \frac{4}{3}\pi(r)^3 = \frac{4}{3}\pi(1 - r^3)$$

We want to see $V_2 = \frac{1}{2}V_1$. So,

$$\frac{4}{3}\pi(1 - r^3) = \frac{1}{2}\frac{4}{3}\pi$$

$$2 - 2r^3 = 1$$

$$r^3 = 1/2$$

Where the real solution is

$$r = (\frac{1}{2})^{\frac{1}{3}} \approx .7937.$$

This means that to double the density, we would have the second ball maintain its mass while being pushed to the $r = .79$. A slight round up and wee see that 80% of the radius hollowed out within to halve the volume.

Similar calculations to see the density quadruple would give you

$$r = (\frac{3}{4})^{1/3} \approx .9086$$

This means that mass in space is very empty to begin with. This model is not changing the density of stars or planets to that of a neutron star (they are much denser, by an extremely large factor of 2×10^{16}).

Simply stated: if this model is correct, then the average density of planets, suns, and systems would be corrected by a small factor of 2 or 4. Definitely less than 10. Not unreasonable.

Density and Earth

In this section, we return to our seismic model where we saw the outside M_1-region and inside m_β-region, shown in Figure 25. Suppose the density of the m_β-region is σ_β and the density of the M_1-region is σ_1. The claim is that σ_1 is denser than σ_β. This implies that

$$1 < \frac{\sigma_1}{\sigma_\beta}.$$

We know that $\sigma_1 = M_1/V_1$ and $\sigma_\beta = m_\beta/V_\beta$. This in turn gives us

$$1 < \frac{M_1}{V_1}\frac{V_\beta}{m_\beta}.$$

Earlier we saw that $M_1/m_\beta = 1/2$. Putting this in we get

$$1 < \frac{1}{2}\frac{V_\beta}{V_1} \implies 2 < \frac{V_\beta}{V_1}.$$

This gives us an idea of how much bigger the volume for m_β must be larger than for M_1. This claim may seem a bit out of context, but what it does is set a relationship with the volumes of the regions based on the ratio of the masses. In construction of this model, this inequality must also be satisfied. For example, we can write spherical volume for M_1 as

$$V_1 = \frac{4}{3}\pi(6.371^3 \times 10^{18} - r^3).$$

The volume for m_β as

$$V_\beta = \frac{4}{3}\pi(r^3 - s^3)$$

where $0 < s < r < 6.371 \times 10^6$, measured in meters.

If

$$r = 6.122 \times 10^6 \text{ and } s = 5.371 \times 10^6,$$

then we get

$$\frac{V_\beta}{V_1} = \frac{6.121^3 - 5.371^3}{6.371^3 - 6.121^3} \approx \frac{74}{29}.$$

And this value is clearly bigger than 2.

<u>The Cavendish Experiment</u>

How are the masses of planets and suns measured to begin with? Mass is measured **indirectly** using Newton's Formulas. By experimentation, we achieved an accurate value for the gravitational constant, G. But this theory does not change these calculations for the mass of planets and stars. Furthermore, it does not contradict (entirely) our current understanding of physics.

Recurrent Nova

There is also this phenomenon of some stars: they seem to nova multiple times. These are known as recurrent nova. There are different explanations we attribute to recurrent novas, one of the main reasons being that stars exist in binary relationships. In a binary system, one star can feed off the mass of the other companion star to nova periodically.

However, there are many known recurrent novas without a binary system. And so the question remained: how can we explain the observation? With this model, we see that a system has much more mass than measured with gravity, and could possibly explain this phenomena as a result of interactions of the star with its "inner" mass.

Cosmic Filaments

As was presented in Figure 15, we saw that such a structure is only stable when large amounts of mass are available. Such structures, would therefore be extremely large, long, and massive if they exist. And they would also have a spin. Are there any objects that satisfy this description?

Indeed, cosmic filaments are definitely in agreement with this. This past month it was published that these cosmic filaments also spin. Their major and minor axis are almost the same, meaning they spin almost in circles.

Cosmic filaments are described as huge bridges of galaxies and dark matter that connect clusters of galaxies and space to each other.

(*Large-Scale asymmetry in galaxy spin directions: evidence from the Southern Hemisphere,* Publications of the Astronomical Society of Australia, Lior Shamir, 14 June 2021: https://arxiv.org/pdf/2106.07118.pdf)

* * *

And lastly, by the definition of COM, it would be impossible to have mass in a **0*D*-space(i.e a point)**. The reason is that a both mass and the COM would have to exist in that same space. The issue being that the COM has a gravity of 0. Newton's Equations don't apply in this space.

General Remarks

Many of the observations I presented involve objects that lack one of the two fundamental aspects to a planetary (and larger) object: enough mass to have noticeable effects of gravity to itself and a constant(ish) rotation. Therefore it makes sense to investigate this system as we have.

LIMITATIONS

This model is not without its limitations in how I presented it:

—By assuming inelastic collisions, we eliminated things like friction and heat. By using elastic collisions, things are more complex.

—The approach is purely theoretical, the lack of planets and stars that have a visible internal structure makes finding empirical evidence difficult. I believe more work can be done to show this a proper model for our universe.

—The Space-Time fabric is better explained with general relativity, but we focus only on objects where we can get away with classical mechanics.

—We do not explore how a magnetic field arises. But the hopes are that with a better understanding of the structure, a proper model for the magnetic field can be made.

—The drawings are not entirely accurate in measurement and size, they are only used as a way to present the idea.

—I wanted to also pose a couple more questions that are unanswered here, but I believe their answers to be consistent with this model:

1) If we consider elastic collisions, which will now factor in friction and heat, how will this alter the model? More specifically, what would it imply about the mass and the kinetic energy of the systems in question?

2) Does the construction of the m_β mass being pushed "apart" while the mass M_1 being pulled together have anything to do with the behavior of the gravity of the masses? Could this be tied to the phenomena of jet streams from the poles of supermassive black holes, quasars, and neutron stars?

CONCLUSION

We started this theory with our fictional *utopium* ball of matter. Ridley's description of general physics inspired that approach in this book. I introduced the idea that rotation allows COM's in higher dimensions to be expressed. We saw how satellites and planets could be linked in their birth. Beyond Newtonian physics, the model could have more insights to stars and black holes, insight to their structures, and perhaps also to the process of fusion. The Zero-Point Paradox, like all paradoxes, is explained upon closer inspection. The evidence I have shown for my model of the spheres is pretty convincing, more so than for the current model. But as a complete theory, this is far from finished. My goal was to illustrate the model as simply as possible.

In summary, by modeling a 3*D* object as a 0*D* object the way we currently do, we may lose a lot of information about the system. We can still make much use of the 0*D* object: to chart trajectories and relationships to other objects at a distance. But if we consider that 3*D* space has a relationship with its 2*D*, 1*D*, and 0*D* space counterparts (namely, by the

COM's), then perhaps we have found explanations for many observations that we currently see.

Lastly, a picture to summarize the difference between the new proposed model and the old model of a rotating system of objects in *3D* space:

| Figure 26 |

New model

Made in the USA
Monee, IL
04 November 2026

67380046R00036